Hadda Ouguissi

Secure signal transmission

Hadda Ouguissi

Secure signal transmission

Secure audio signal transmission using chaotic systems

ScienciaScripts

Imprint

Any brand names and product names mentioned in this book are subject to trademark, brand or patent protection and are trademarks or registered trademarks of their respective holders. The use of brand names, product names, common names, trade names, product descriptions etc. even without a particular marking in this work is in no way to be construed to mean that such names may be regarded as unrestricted in respect of trademark and brand protection legislation and could thus be used by anyone.

Cover image: www.ingimage.com

This book is a translation from the original published under ISBN 978-620-6-70640-3.

Publisher:
Sciencia Scripts
is a trademark of
Dodo Books Indian Ocean Ltd. and OmniScriptum S.R.L publishing group

120 High Road, East Finchley, London, N2 9ED, United Kingdom
Str. Armeneasca 28/1, office 1, Chisinau MD-2012, Republic of Moldova, Europe
Printed at: see last page
ISBN: 978-620-7-71118-5

Contents

Ouguissi Hadda

Secure transmission of an audio signal using chaotic chaotic systems

- Mixing two chaotic maps (Logistics map and Tent map)
- The chaotic Chua system

Voice communication is closely linked to everyday life, such as education, e-learning, commerce, politics and the dissemination of information. In order to maintain security, sensitive data must be protected during transmission.

Chaotic systems have become good candidates for cryptography to increase the degree of security.

In this book, we focus on the design of a high-security voice communication system using two levels of encryption based on chaotic systems. The first level is chaotic masking, while the second level is scrambling. In the masking level using two methods, the first method is masking using a hybrid between the chaotic frameworks (logistic map and tent map), while the second method is based on the Chua chaotic system. At the scrambling level we use Amold's map. At the receiver level, we use Pecore and Carroll synchronisation to recover the encrypted audio signal.

Ouguissi Hadda is a telecommunications specialist who obtained her doctorate in electrical engineering from Amar Thelidji University in Laghouat, Algeria, in 2022. My research focuses on secure data transmission using chaotic systems.

Voice communication is closely linked to everyday life, such as education, e-learning, commerce, politics and the dissemination of information. To maintain security, sensitive data must be protected during transmission.

Chaotic systems have become good candidates for cryptography to increase the degree of security.

In this book, we focus on the design of a high-security voice communication system using two levels of encryption based on chaotic systems. The first level is chaotic masking, while the second level is scrambling. In the masking level using two methods, the first method is masking using a hybrid between the chaotic frameworks (logistic map and tent map), while the second method is based on the Chua chaotic system. At the scrambling level we use Amold's map. At the receiver level, we use Pecore and Carroll synchronisation to recover the encrypted audio signal.

Ouguissi Hadda is a telecommunications specialist who obtained her doctorate in electrical engineering from Amar Thelidji University in Laghouat, Algeria, in 2022. My research focuses on secure data transmission using chaotic systems.

Introduction generate

General introduction

Voice communication is closely linked to everyday life, such as education, e-learning, commerce, politics and the dissemination of information. With the development of modern communication and multimedia technologies, a huge amount of sensitive voice data travels daily over open and shared networks. In order to maintain security, sensitive data must be protected during transmission.

Researchers have proposed a large number of ways of encrypting a speech signal, just as conventional cryptographic techniques are effective for text data [1], [2].

One of the potential solutions that has an association with the growth of non-linear communication systems and voluminous, redundant voice data is chaos. Thanks to the properties of chaotic systems, such as the high sensitivity of non-linear systems to initial conditions, and the fact that they evolve in a wide frequency band, which makes their trajectories appear as pseudo-random noise, chaotic systems have become good candidates for cryptography in order to increase the degree of security [3-6].

Chaotic systems have been applied to cryptography in order to increase the degree of security over the last few decades. The systems remained unknown until the 20th century. Henri Poincare [7] discovered the notion of sensitivity to initial conditions through the problem of the interaction of three celestial bodies. Later, in 1960, the work of Edward Lorenz [8], who was passionate about meteorology, set the course for this branch of mathematics.

Encryption and decipherment techniques using a low-dimensional chaotic system have a small key space, these techniques offer low resilience against brute force attacks [9]. Many random encryption techniques take full advantage of the integration of many chaotic systems to optimise the key space, but this leads to higher computational complexity and computation time [10]. Encryption techniques based on high-dimensional chaotic systems that exhibit very complex behaviour have been proposed [11-12].

Thanks to speech scrambling techniques [13], combining both chaotic scrambling and voice masking encryption can produce very low intelligibility and high encryption strength [14]. The development of communication systems using chaos began with very simple synchronization schemes of electronic circuits, aimed at simultaneous encryption and reconstruction of an information signal [15-17].

Furthermore, the 1 information signal may itself be a chaotic signal, even if the chaotic signal is originally unpredictable, it will be controlled so that it does not change its behaviour but carries a useful signal.

In this book, we focus on the design of a high-security voice communication system using two levels of encryption based on chaotic systems. The first level is chaotic masking, while the second level is scrambling. In the masking level using two methods, the first method is masking using a hybridisation between the chaotic frameworks (*logisticsmap* and *tent map*), while the second method is based on the *Chua* chaotic system. At the scrambling level we use Arnold's map. At the receiver level, we use Pecore and Carroll synchronisation to recover the encrypted audio signal.

This book is organised as follows:

Chapter 1 is devoted to an overview of chaotic systems, followed by some examples of chaotic systems and their characteristics.

In Chapter 2, we propose techniques for encrypting audio (scrambling), and present methods for masking a signal, before briefly introducing the concept of digital watermarking.

In Chapter 3, we present a method for synchronising the two chaotic systems.

In Chapter 04, we discuss the methods of chaotic masking and chaotic scrambling, and present the method of synchronising the two *Chua* chaotic systems by *Pecora and Carroll*.

Finally, we end this modest work with a conclusion.

Chapter 1

General information on chaotic systems

1.1 Introduction

The aim of this chapter is to present some generalities about chaotic systems, then we give some examples of chaotic systems, and then we study the *Chua* chaotic system and its control method.

1.2 Generalities on chaotic systems

The term "chaos" defines a particular state of a system whose behaviour never repeats itself.

There are several possible definitions of chaos [18-19]. These definitions are not all equivalent, but they converge towards certain common points that characterise chaos:

> Non-linearity: if the system is linear, it cannot be chaotic.

Determinism: a chaotic system has fundamental rules that are determinist (rather than probabilistic).

> Sensitivity to initial conditions: very small changes to the initial state can lead to radically unpredictable behaviour.

> Imprevisibility: because of the sensitivity to initial conditions, the response is totally unpredictable after a certain time of evolution.

Irregularity: hidden order comprising an infinite number of unstable periodic patterns (or motions). This hidden order forms the infrastructure of chaotic systems.

1.2.1 Characteristics of chaos

In what follows, we present a number of characteristics that provide a qualitative understanding of the key points of a chaotic system:

1.2.1.1 Sensitivity to initial conditions

Chaotic systems are extremely sensitive to disturbances. This concept is illustrated by the famous "butterfly effect", described and popularised by the meteorologist Edward Lorenz [8]. The evolution of a chaotic dynamic system is unpredictable in the sense that it is sensitive to initial conditions. Thus, two initially similar phase trajectories always diverge from each other. We will illustrate this phenomenon with a numerical simulation of the *logistic map* equation.

1.2.1.2 Strange attractors

An attractor is the zone in phase space that attracts the trajectories of a dynamical system. The simplest attractor is a point. There are two types of attractors, regular attractors and strange or chaotic attractors. In the case of a chaotic system, the trajectory converges towards a particular region of space called a strange attractor, which is a sign of chaos [8].

1.2.1.3 Lyapunov exponent of a chaotic system

The Lyapunov exponent is used to measure the degree of stability of a system, a system sensitive to very small variations in initial conditions will have a positive exponent (chaotic system).a strange attractor will always have at least one positive Lyapunov exponent, in other words the largest exponent is positive for a chaotic system and negative for other systems [20], [21].

1.3 Some examples of chaotic systems

There are several chaotic systems that are used to generate chaotic signals. In this section, we present two classes: continuous chaotic systems and discrete chaotic systems.

1.3.1 Discrete chaotic systems :

-The logistics card (one dimension only):

The logistic chaotic function is expressed as :

$$x_{n+1} = rx_n(1 - x_n) \tag{1.1}$$

Or x_0 takes a value in the interval [0,1] called the initial conditions.

r is a positive constant and takes a value from 0 to 4 as shown in Fig.1.1 and Fig.1.2.

To illustrate the behaviour of a dynamic system, let's take as an example the logistic map described by equation (1.1).

For each value of $r \epsilon$ **[0,4]** we have chosen a sequence of **500** samples with a transition period of **500** samples. Depending on the value of r (the bifurcation parameter) and the initial value of x_0 of the sequence x_k , the sequence exhibits very different behaviours..:

♦   ♦♦ for **r = 2.7** and x_0 = **0.15** Involution of sequence x_k converges rapidly to a fixed and stable point on the plane (x_k, x_{k+1}).

♦   ♦♦ for r = 3.2 we see that the sequence converges to a periodic solution. In this case, the trajectory converges to a cycle of order 2.

♦   ♦♦ By increasing the value of r the new sequence converges to a periodic solution with a doubling of the period.

♦   ♦♦ For $r \geq 3,57$ the following x_k no longer represents an ordered structure. Done, the system becomes chaotic as shown in Fig.1.2.

The road to chaos can be summarised using the bifurcation diagram given by Figl.I

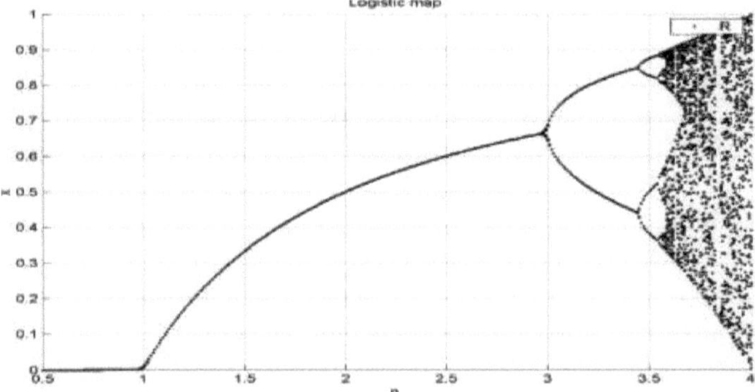

Fig.1.1 - Logistic map bifurcation diagram

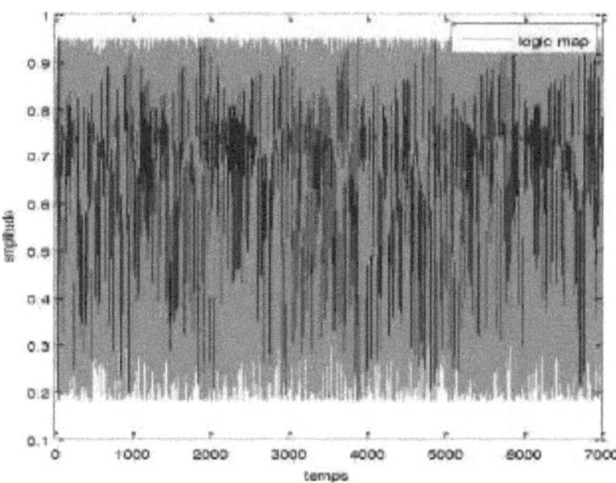

Fig.1.2-The chaotic signal of the logistic map.

This logistic map is very sensitive to initial conditions. We will set two very close initial conditions $x_0 = $ **0.25** and$_0 = $ **0.251** for the chaotic system (1.1). Initially, the two systems evolve in the same way, but very quickly, their behaviour becomes different as shown in Fig.1.3.

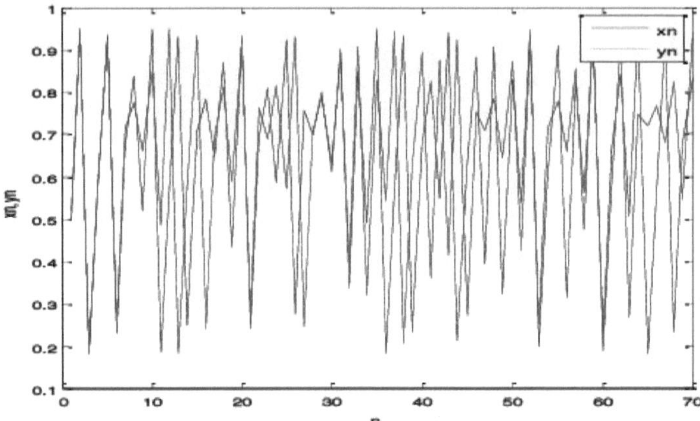

Fig.1.3 -Devolution over time for two very similar initial conditions.

-The *tent map*

Is a piecewise linear map, described by the following equation:

$$\begin{cases} f(x_i, u) = ux_i & si \quad x_i < 0.5 \\ f(x_i, u) = u(1 - x_i) & si \ autrement \end{cases} \qquad (1.2)$$

Ou $x_i \epsilon [0,1]$ pour $i \geq 0$.

And u is the control parameter which varies in the interval [0,2], x_0 is an initial system value.

Depending on the control parameter u, the system illustrates a variety of dynamic actions ranging from expected to chaotic as shown in Fig.1.4.

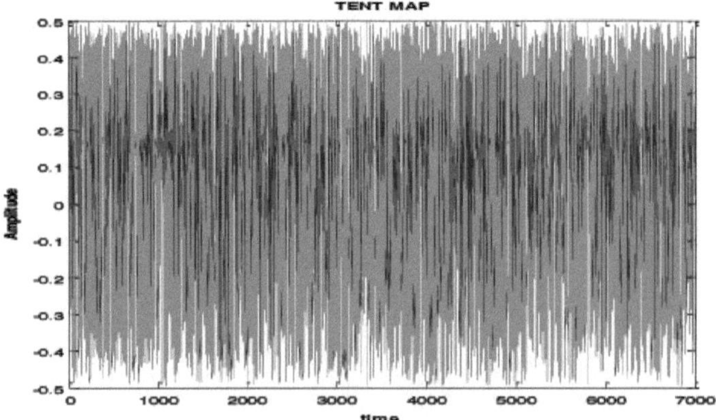

Fig.1.4-the chaotic signal of the tant map.

-The ARNOLD card

The chaotic map called Arnold's map in recognition of the Russian mathematician Vladimir I. Arnold, who discovered it using an image of a cat. It is a simple and elegant demonstration and illustration of certain principles of chaos, an apparently random evolution of a system.

If we consider $X = \begin{pmatrix} x \\ y \end{pmatrix}$, a matrix of size X N ,1the Arnold transformation T is:

$$\begin{pmatrix} \dot{x} \\ \dot{y} \end{pmatrix} \equiv \begin{pmatrix} 1+1 \\ 1+2 \end{pmatrix} \begin{pmatrix} x \\ y \end{pmatrix} mod\ N \qquad (1.3)$$

Where' mod N is modulo N, (x, y) are the coordinates of the original watermark and (x', y') are the coordinates of the scrambled watermark. N is the height or size of the signal to be processed.

1.3.2 Continuous chaotic systems :

-Lorenz system

In 1963, Edward Lorenz modelled a differential system with chaotic behaviour for certain parameter values. This system is defined by the following equations:

$$\begin{cases} \dot{x} = \sigma(y - x) \\ \dot{y} = -rx - y - xz \\ \dot{z} = -bz + xy \end{cases} \qquad (1.4)$$

Below is the Lorenz attractor for the following values $\sigma = 10, r = \frac{3}{8}, b = 28$,and initial conditions $x_0 = y_0 = z_0 = 0.01$ with a simulation step of **0.01.**

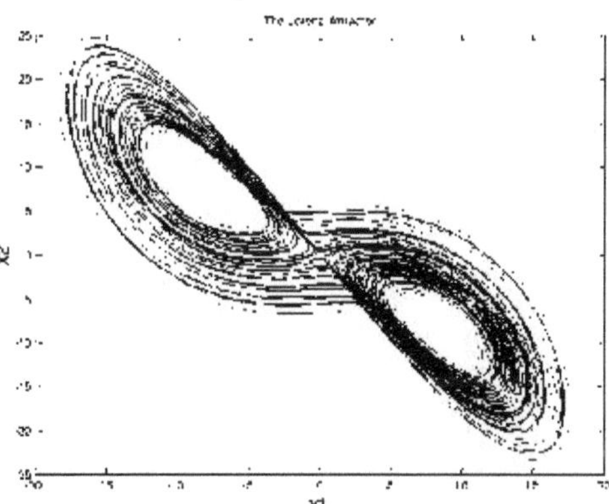

Fig.1.5 Lorenz chaotic system

- *Chua* circuit

The *Chua* circuit is a simple electronic circuit (Fig.1.6) which exhibits classical chaos theory behaviour [22-24]. It was presented in 1983 by **Leon O. Chua,** who was a visitor at Waseda University in Japan at the time.

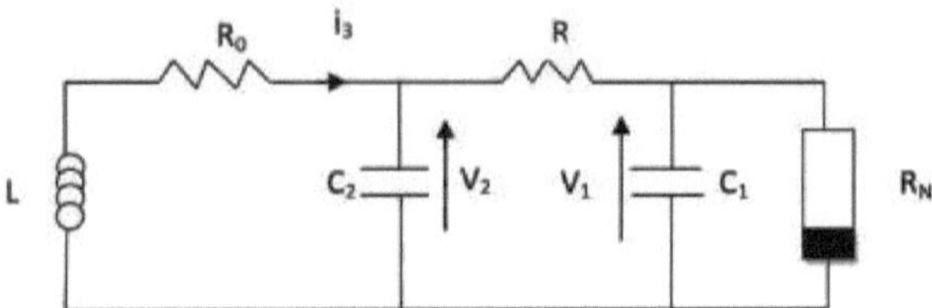

Fig.1.6 -*Chua* circuit.

The *Chua* system represented by the set of differential equations :

$$\begin{cases} \dot{x} = \alpha(y - x - f(x)) \\ \quad \dot{y} = x - y + z \\ \quad \dot{z} = -\beta(x - R_0 z) \end{cases} \qquad (1.5)$$

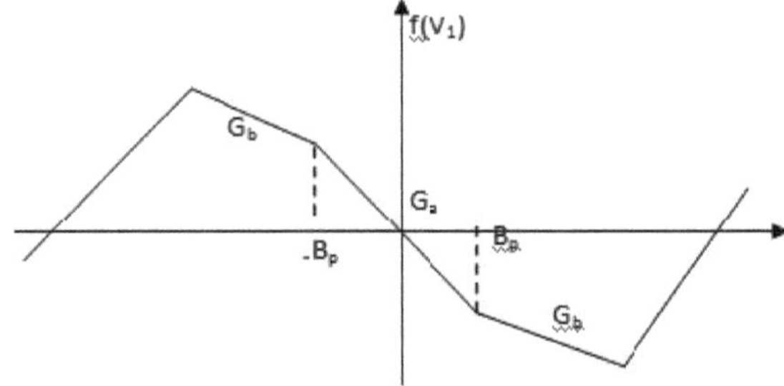

Fig.1.7 - Voltage-current characteristic of the non-linear resistor.

$$\text{Ou' } f(x) = m_1 x + \frac{m_0 - m_1}{2}(|x + 1| - |x - 1|) \qquad (1.6)$$

The parameters of this circuit depend essentially on the values of the resistance, the inductance and the capacitors:

$$\alpha = \frac{C_2}{C_1} , \beta = \frac{R^2 C_2}{L} , m_0 = \frac{G_a}{G} , m_1 = \frac{G_b}{G} \qquad (1.7)$$

With $G = {}^{1}\!/_{R}$

1.4 Chaotic system control :

In order to conceal a confidential message by superimposing it on a chaotic system, the information signal must itself be a chaotic signal, but as a chaotic signal is inherently unpredictable, it is therefore necessary to control it so that it does not change its behaviour but carries an information signal.

From the above, we have seen that we can reach the chaotic regime by varying the control parameter.

1.4.1 Chaos control technique

It is clear that the dynamic behaviour of a non-linear system can be changed by changing some of its parameter values. In chaos control, we have to work in phase space, parameter space and the Poincare map. In addition, Lyapunov exponents and the bifurcation diagram are typical tools for the study.

Chaos control is achieved by stabilising the unstable periodic orbits of a chaotic system by applying a small perturbation to certain parameters of the system, or the first to evoke this notation Ott, Gerbogi and York (OGY) [25], and because it has become possible to control a chaotic system that is evaluated chaos control in electronic circuits [26].

1.4.2 Control of the *Chua* chaotic system

Chua system control based on Lyapunov exponents and the bifurcation diagram.

We have the *Chua* system (1.5), the fixed points of this system are:

$$C_0(0,0,0), C_1\left(\frac{m_0 - m_1}{m_1 + 1}, 0, \frac{m_1 - m_0}{m_1 + 1}\right), C_2\left(\frac{m_1 - m_0}{m_1 + 1}, 0, \frac{m_0 - m_1}{m_1 + 1}\right)$$

1.4.2.1 Chaotic system behaviour of *Chua*

It is sufficient to change the value of a component to study the various modes or behaviours it can exhibit in Chua's circuit. In Chua's circuit, we take the resistance a as the bifurcation parameter, we have carried out simulations by varying the parameter a from **10** to **17** with an iteration step of **0.01,** and et $\beta = 28, m_0 = -1.27, m_1 = -0.68$.

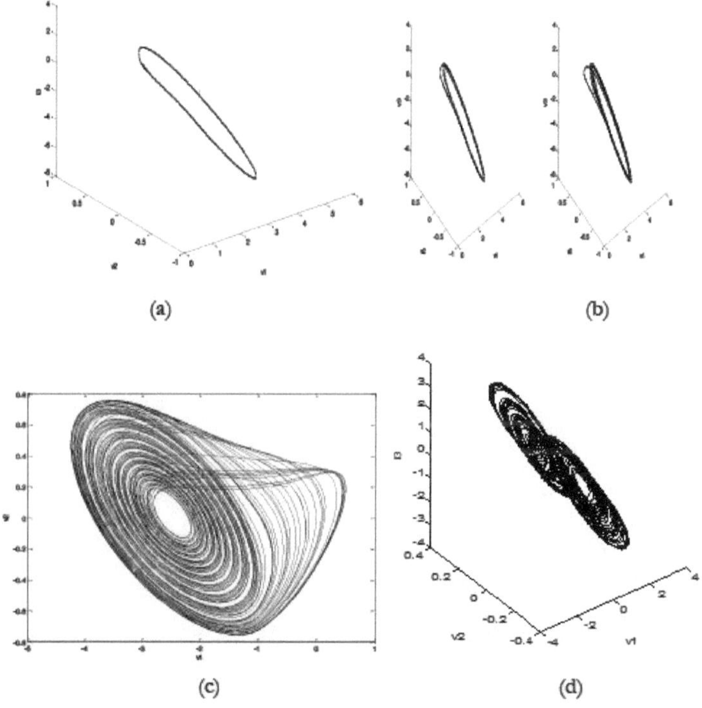

(a)   (b)

(c)   (d)

Fig.1.8-The behaviour of the *Chua* circuit for different values of a

1.4.2.2 Stability by Lyapunov exponent:

We take the values of the parameters of the *Chua* system (1.5) as the chaotic case

$$\alpha = 15.6, \beta = 28 \; m_0 = -1.27, m_1 = -0.68,$$, we choose the initial values of the state

as $x_0 = 1, y_0 = 0.5, z_0 = -1$ the Lyapunov exponent of the system (1.5) are:

$$L_1 = 0.2, L_2 = 0, L_3 = -4.3,$$ the system (1.5) is chaotic because it has a positive Lyapunov

exponent L_1.

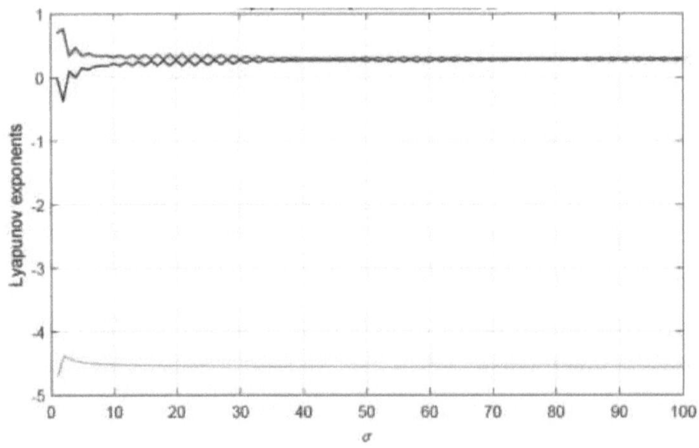

Fig.1.9.Lyapunov exponent of the Chua system.

1.5 Conclusion

In this chapter, after presenting some examples of chaotic systems and their characteristics, we have found that it is possible to control non-linear systems by controlling their parameters. We have found that it is possible to control non-linear systems by controlling their parameters. Some application of chaos control, which consists of designing a controller to allow the transmission of information in a communication system.

Chapter 2

Secure transmission of a chaos-based audio signal

2.1 Introduction

Chaotic sequences have properties similar to the cryptographic properties of confusion and diffusion, and have therefore been used to build good cryptosystems. These properties make chaotic cryptosystems resistant to statistical attacks [27]. Also, the use of chaos in the cryptosystem will ensure security, complexity, speed and computational power [28].

2.2 Cryptography

Cryptography is the study of methods for transmitting data confidentially. In the secure transmission of information, the message, known as the text, is transformed in such a way as to make it incomprehensible; this process is known as "encryption". The recipient must also carry out a process called "deciphering" or "decrypting", to reconstruct the message from the ciphertext.

In standard cryptography, and among a wide variety of encryption mechanisms, there are two types of key [29]: secret key and public key. In a secret-key algorithm, the encryption key is calculated from the decryption key and vice versa. In general, the encryption and decryption keys are identical. In public key encryption or asymmetric encryption, the encryption key is different from the decryption key. Anyone can use the encryption key to encrypt a message, while the private key is used to decrypt a message, and the two keys are linked. Public key encryption may be preferred for generating small sequences, symmetric encryption may be preferred for encrypting large quantities of data.

2.3 The chaotic encryption of audio

Speech is recorded in the form of an analogue signal which is converted by the sound card into digital by sampling the signal for storage in the computer.

2.3.1 The characteristics of audio

A simple definition of sound can be given to its waves resulting from a change in atmospheric pressure, and although this change does not exceed (±1) but when it comes into contact with the inner ear of the human auditory system it operates in a wide dynamic manner at frequencies between 20Hz-20KHz. Sound is represented by a continuous line diagram known as the wave and pitch represent the amplitude of sound (volume), and this formula is called the analogue signal as illustrated in Fig.2.1(a).

The computer and certain electrical devices treat things as a series of binary numbers (0,1), i.e. in digital form, and for this it was necessary to find a way or method of converting the sound from its analogue state to the digital signal form so that this device could understand it and treat it as desired.

2.3.1.1 Digital audio recording:

When sound is stored in the computer via a sound outlet connected to the computer's sound card, the sound outlet (microphone) converts the voltage fluctuations into the form of an analogue signal. The signal is measured and converted into parts called samples [30], then converted into a series of numbers by a process called quantisation [31], then converted into binary form and stored. These samples are stored inside the computer on the secondary memory (hard disk) in binary digital format. An electronic circuit on the sound card called an ADC (analogue to digital converter) helps with these operations. When the audio is played back, the process is reversed, but the fluctuation in voltage will be transmitted to the loudspeakers instead of the sound pick-up, and then converted into a fluctuation in atmospheric pressure, there is also an electronic circuit that restores the sound to its analogue state again called a DAC [32-33], as shown in Fig.2.2, certain factors affect the digital sound recording process, including:

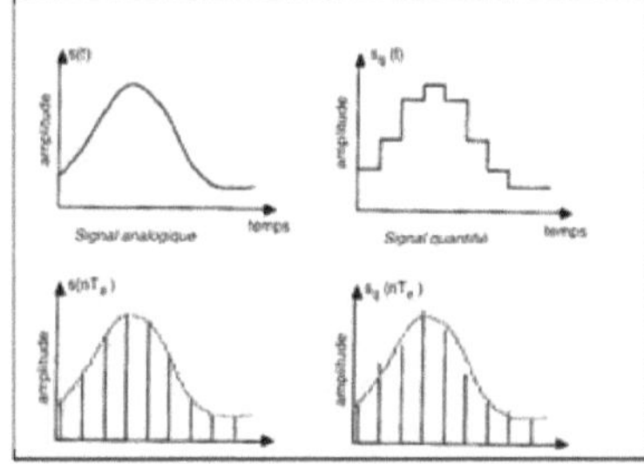
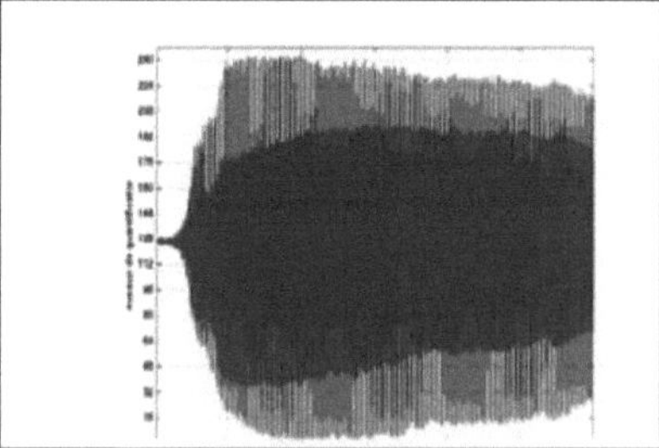

Fig.2.1 (a)-Morphological classification of signals (b)-Signal quantised on 8 bits.

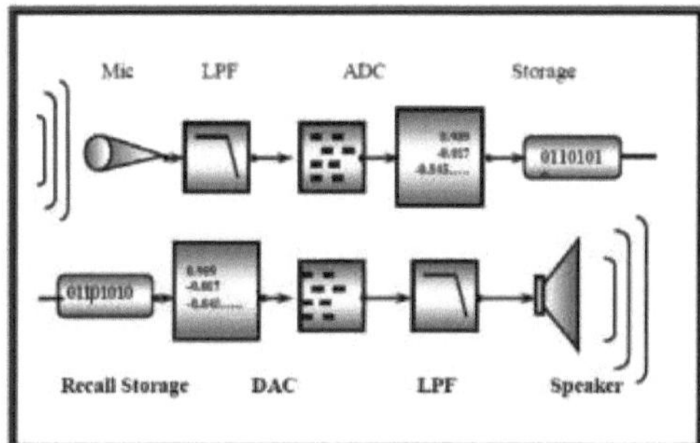

Fig.2.2 digital audio recording process

The sound is stored in different files depending on the different storage formats, such as files with an mp3 extension and a wav extension.

2.3.1.2 Audio file with extension (wav):

WAV audio files are one of the audio formats used by Microsoft in the Windows system environment, and are one of the most popular and widely used formats in the file environment defining the (common) RIFF (Resource Interchange File Format) file format. RIFF files are organised into overlapping and interconnected segments that include a definition of the RIFF content [34].

2.3.2 Speech encryption techniques

As voice communications become more widely used and even more vulnerable, the importance of ensuring a high level of security is a major issue. To date, numerous speech encryption techniques have been proposed. Speech encryption techniques can be classified into four types: Temporal domain scrambling, frequency domain scrambling, amplitude scrambling and two-dimensional mixed scrambling [1], [2]. Chaotic signals are used and by means of a permutation key produced by a chaotic generator, the original voice signal where encrypted in the transmitter and recovered in the receiver, with the initial conditions and parameters of these chaotic maps, is identical in transmitter and receiver [35].

2.3.2.1 Interference in the frequency domain:

Frequency-domain signal scrambling is performed by applying a random permutation key to the discrete

Fourier transform (DFT) of the speech signal and to the half-range (conjugate symmetry) of the frequency component. The signal spectrum is divided into sub-bands, and these sub-bands are permuted by a permutation key generated by a random number generator [35].see Fig.2.3.

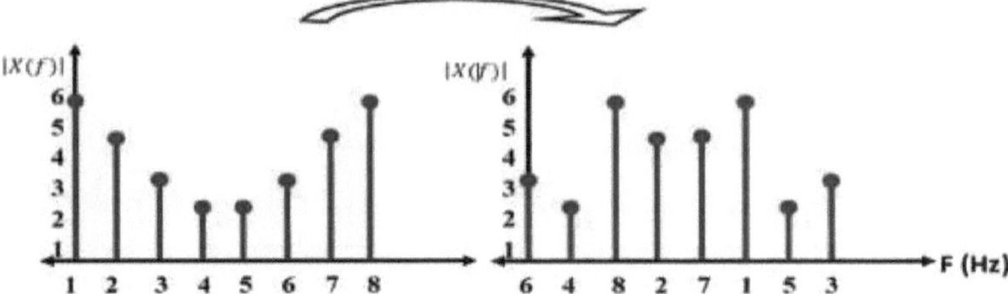

Fig.2.3. Interference in the frequency domain :

2.3.2.2 Interference in the time domain:

It divides the speech signal into frames, and each frame is divided by time domain into segments (sub-frames) and then scrambled by a permutation key produced by a chaotic generator as illustrated by the following example:

We use the logistic chaotic map (1.1) defined in the previous chapter. A chaotic sequence is generated with a length equal to the length of the term L.

Let **N=9,** and with the initial condition $x_0 = 0.25,$ and the bifurcation parameter $r = 3.9$ the generated sequence value is:

(a)

i	1	2	3	4	5	6	7	8	9
x_i	0.5000	0.9500	0.1805	0.5621	0.9353	0.2298	0.6726	0.8369	0.5188

The previous chaotic sequences were drawn using the ascending order algorithm:

(b)

i'	3	6	1	9	4	7	8	5	2
x'_i	0.1805	0.2298	0.5000	0.5188	0.5621	0.6726	0.8369	0.9353	0.9500

After drawing the chaotic vectors in step 2, the indices of the elements are modified. These indices represent the new index for scrambling in the transmitter and the receiver for encrypting this audio, for our sequence in step 2, the new representation of the index is:

(c)

Indice d'entrée	1	2	3	4	5	6	7	8	9
Indice de sortie	3	6	1	9	4	7	8	5	2

2.3.2.3 Two-way interference:

Which combines frequency and time scrambling [35].

2.3.2.4 Amplitude interference:

Also known as a masking technique in which the speech signal is converted by pseudo-random or chaotic amplitudes.

2.4 Chaotic masking methods

Chaotic signals are used as an information carrier. To do this, the message is encrypted by the transmitter and decrypted and extracted from the chaotic signal by the receiver. In the communication domain, information retrieval is generally based on synchronisation between the transmitter and the receiver [36]. There are several methods of injecting information into a chaotic system. Among the methods of chaotic transmission, we can mention masking by addition [37], encryption by inclusion [38], and encryption by modulation [39], etc.

2.4.1 Chaotic masking by addition

This technique, developed in 1993, is known as chaotic masking [37]. The principle of this scheme is to perform a simple addition between the transmitter output signal y(t) and the message m(t). The sum of the two signals is transmitted to the receiver over a public channel. The receiver consists of a system identical to that of the transmitter, a simple subtractor. Once the two chaotic systems (transmitter and receiver) have been synchronised, the message is extracted using a subtraction operation. The diagram representing this method is given in Fig.2.4.

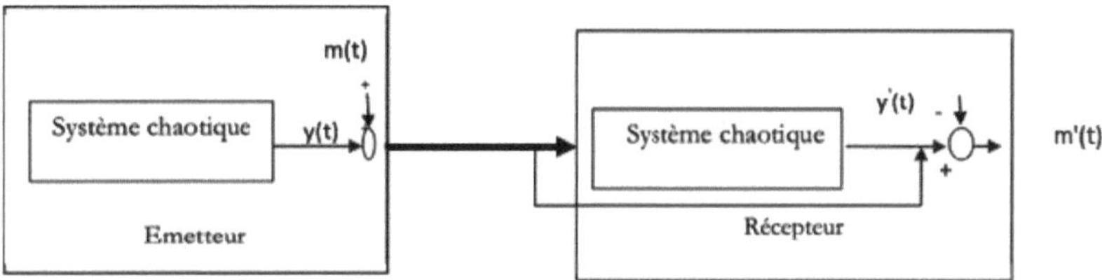

Fig.2.4-chaotic addition masking

2.4.2 Masking by parametric modulations

The principle of this technique is to use the message m(t) to modulate one of the parameters of the transmitter's chaotic system. A sliding-mode observer is loaded to ensure synchronisation at the receiver.

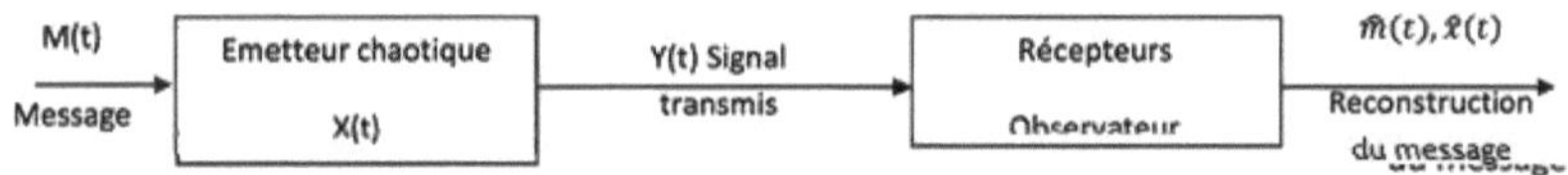

Fig.2.6. masking by chaotic modulation.

2.4.3 Inclusion masking

This encryption technique consists of injecting the message into the transmitter's dynamics [38] with parameter modulation [39] and, using a sliding-mode observer, the information is restored.

And to prove the success of the proposed encryption system we will add a watermark to the original signal during the encryption process and extract this watermark during the decryption process. Successful extraction of the watermark during the decryption process confirms that the signal received is authenticated and therefore not subject to any attacks.

2.5 Digital tattooing

Digital watermarking is a processor that consists of inserting a digital mark (random sequence, binary logo, etc.) into an original host signal (image, text document, sound, etc.) in an imperceptible and indelible way. This mark contains data that can be used in a variety of applications, including copyright protection, transmission monitoring, data authentication or secure transmission. In the case of audio watermarking, various techniques have been used to apply the watermark such as transform techniques (DWT, DCT), algebraic techniques [40-41], a new blind watermark design for speech signals, and have used discrete wavelet transform (DWT) and discrete cosine transform (DCT) after signal segmentation [42], a robust blind audio watermarking method using DCT and DWT within signal subsampling [43-44]. To achieve high quality imperceptibility, fusion is performed against various attacks such as: re-quantization, cropping, echo, amplification and additive white Gaussian noise (AWGN). In Appendix B details the DWT, DCT watermarking techniques.

2.6 Security analysis

2.6.1 Cryptanalysis of chaos-based communication systems

Cryptanalysis is the study of the probability of success of possible attacks [45] on cryptosystems in order to detect any weaknesses. A cryptosystem must have two fundamental cryptographic properties: confusion and diffusion. Confusion consists in making the relationship between the key and the encrypted message as complex as possible. Diffusion means that a small change in the plaintext message or in the key induces a large change in the cipher message. Cryptanalysis aims to find the key space and the sensitivity.

2.6.1.1 Analysis 1'space of the secret key

The security of a communication system must depend solely on its secret key and, according to Kirchhoff's principle [46], the size of the K key space is defined by the number of key pairs used in encryption and decryption, the total number of different keys used in the encryption system briefly referred to as the key space [47]. In addition, a good encryption system must have a large key space to compensate for degradation in the PC, and thus prevent invaders from decrypting the original data even after investing large amounts of resources and time [48]. According to [49] the design of a cryptosystem that resists brute force attacks, the size of the key space must be greater than $2^{128} (\approx 3.24 \times 10^{38})$.

2.6.1.2 Key sensitivity analysis

Key sensitivity analysis is extremely important when designing a secure transmission scheme. Changing one bit in a key generates a completely different cipher text. Calculating the Unified Average Variable Intensity (UACI) and Number of Sample Change Rates (NSCR) between the two encrypted speech signals to assess the sensitivity of the key. The NSCR and UACI of the two encrypted speech signals are calculated using the equation below [50].

$$\text{NSCR} = \sum_{i=1}^{L} \frac{d_i}{L} \times 100\% \qquad ou\,'d_i = \begin{cases} 1 & S'_{1,i} = S'_{2,i} \\ 0 & auterment \end{cases} \qquad (2.1)$$

$$\text{UACI} = \frac{1}{L} \left[\sum_{i=1}^{L} \frac{S'_{1,i} - S'_{2,i}}{Max} \right] \qquad (2.2)$$

Where' S'-y,S'_{2i} i are the two speech signals with a slow difference on the key in I^{Ieme} samples.

L: represents the length of the speech vector.

Max : depends on each voice and audio signal sample assuming an integer value in the range [0-65535] and in this situation Max=65535, done when using the Matlab environment, digital audio are normalized in the range

[-1,1], then the maximum is 2.

2.6.1.3 Encryption quality

The quality of the speech signal extracted from the encrypted signal is an essential element, so we will discuss the quality of the signal extracted from the encrypted signal, using the two coefficients (correlation and SNR).

2.6.1.3.1 Correlation coefficient :

Digital audio is characterised by highly redundant and highly correlated adjacent samples. A robust secure audio transmission scheme must be able to eliminate this type of relationship.

The correlation coefficient is a measure of the linear relationship between two variables [51]. If two variables are closely related with a stronger association, the correlation coefficient is close to the value 1. On the other hand, if the coefficient is close to 0, the two variables are not related and cannot be predicted.

The correlation coefficient "r" can be calculated using the following formulae [52]:

$$r_{S_1 S_2} = \frac{\frac{1}{L}\sum_i^L (S_{1,i} - E(S_i))(S_{2,i} - E(S_2))}{\sqrt{\left(\frac{1}{L}\sum_i^L (S_{1,i} - E(S_i))^2\right)} \times \sqrt{\left(\frac{1}{L}\sum_i^L (S_{2,i} - E(S_2))^2\right)}} \qquad \text{Où } E(s) = \frac{1}{L}\sum_{i=1}^L S_1 \tag{2.3}$$

Where L is the length of the voice signals (number of samples).

$S \, S_{132}$ are the dual nature of the two signals (original, encrypted) or (original, decrypted).

2.6.1.3.2 Signal to noise ratio (SNR) :

To confirm the performance of digital speech encryption schemes, the SNR is calculated, or SNR measures the continuous noise in the encrypted speech signals. Crypta analysts always try to increase the continuous noise in the encrypted signal in order to minimise the information in the encrypted data. The decryptor also tries to reduce the noise continuum in the decrypted signal. The signal-to-noise ratio is a factor used to identify the quality of noise on the signal, the signal-to-noise ratio can be calculated by the equation below [53]:

$$\text{SNR} = 10 \times \log 10 \frac{\sum_{i=1}^L S_{i,1}^2}{\sum_{i=1}^L (S_{1,i} - S_{2,i})^2} \tag{2.4}$$

$S_{1,i}, S_{2,i}$ represent the $\text{I}^{\text{ième}}$ samples of the (original, unencrypted) or (original, unencrypted) speech signals respectively.

L Represents the length of the voice signals.

To authenticate that the received encrypted voice signal was sent from the trusted side, we have the watermark extraction examined by the BER

2.6.2 Bit Error Rate (BER)

The BER is used to check the similarity between the two watermarks, the original and the extracted watermark image. Furthermore, a BER of zero means that there is no effect on the watermark and that the extraction is successful, which means that the received voice signal is sent on the authenticated side. The BER is expressed by the following formula [42] :

$$\text{BER} = \frac{B_{ERR}}{N} \times 100\% \qquad (2.5)$$

Or' B_{ERR} the quality of erroneous bits.

N: the number of all the bits (watermark size).

2.7 Conclusion

In this chapter, we have presented the chaotic audio encryption techniques, the main method based on amplitude scrambling (masking), and time scrambling, and then we have added a watermark where we have given a brief overview of digital audio watermarking. Finally, we have given some performance evaluation measurements on the transmitter side. On the receiver side, in order to reconstruct the masked audio, we need to synchronise the transmitter and the receiver, as we will explain in the next chapter.

Chapter 3

Synchronisation of two systems chaotic

3.1 Introduction

Synchronisation occurs when two dynamic systems evolve in the same way in time. The history of this phenomenon goes back to the 17ᵉᵐᵉ century, when the Dutchman Christian Huygene (1629-1695) made his observation on two clocks with slightly different frequencies. For some years now, the theory of chaotic systems has been applied in the field of communications, and it is difficult to synchronise two chaotic systems in the real world. It is extremely difficult to build two identical circuits because of the tolerance on the components, and the sensitivity of chaotic systems to initial conditions. Researchers have recently discovered solutions for the success of the synchronisation process.

3.2 Synchronising two chaotic systems

Because of the sensitivity of these chaotic systems to initial conditions, their synchronization seems impossible at first. In 1983, the question of synchronisation was addressed using piecewise linear electronic circuits [55]. In the 1990s, *Pecora* and *Carroll* [56-57] showed that two identical chaotic systems with different initial conditions can eventually be synchronised under certain conditions.

3.2.1 Chaotic synchronisation methods

Traditional methods of synchronisation are generally based on the use of identical circuits. Suppose two identical chaotic systems are oscillating totally independently, if by some means they are allowed to exchange energy, an action known as "coupling", the two systems will eventually give way to a common behaviour - they will synchronise.

3.2.1.1 Rated systems:

Chaotic systems can be coupled in one direction (unidirectional coupling) or in both directions (bidirectional coupling) [58].

Bidirectional coupling: in this case, the coupling element allows energy to be exchanged in both directions. In practice, a resistor is used to ensure this type of coupling.

Unidirectional coupling: in the case of unidirectional coupling, the energy is transferred from one system to another using a coupling element operating in one direction only, such as a server. In practice, for example, a simple amplifier-based electronic circuit mounted as a follower performs this task.

To illustrate these two concepts, let's take the two identical chaotic systems a and b of dimension 3 described by $\dot{x}_a = f(x_a) \text{ et } \dot{x}_b = f(x_b)$. The diagram of possible couplings between the two systems is given in Fig.3.1.

(a)

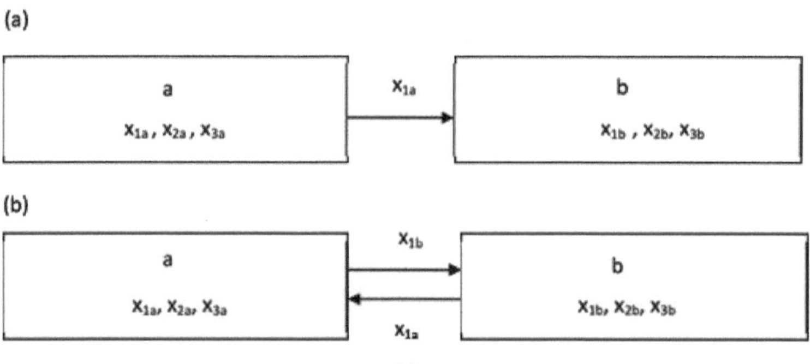

Fig.3.1 : Coupling diagram: (a) unidirectional, (b) bidirectional.

Definition3.1: Synchronisation can be described by the following definition:

Consider the two systems

$$\begin{cases} \dot{x} = f_1(x) \\ \dot{y} = f_2(y) \end{cases} \qquad (3.1)$$

With $x, y \in R^n$, $f_1 et f_2$ two non-linear functions defined by $R^n \to R^n$. The two systems are said to be synchronised if :

$$\lim_{t \to \infty} \|y(t) - x(t)\| = 0 \qquad (3.2)$$

Where $y(t) - x(t)$ represents the synchronisation error for all initial conditions $x(0)$ and $Y(0)$.

3.2.2 Types of synchronisation

There are several types of synchronization proposed in the literature, such as unidirectional coupling synchronization, bidirectional coupling synchronization, *Pecora* and *Caroll* identical synchronization [55-58] and observer synchronization [59].

3.2.2.1 Synchronisation using an observer

The unidirectional synchronisation of two chaotic systems can be considered as a non-linear observer problem, where the use of observers is proposed to estimate unknown states of a system that are not directly measurable. Several types of observer have been reported in the literature, including adaptive observers [60] and sliding mode observers [61]. Fig.3.2 illustrates this synchronisation principle.

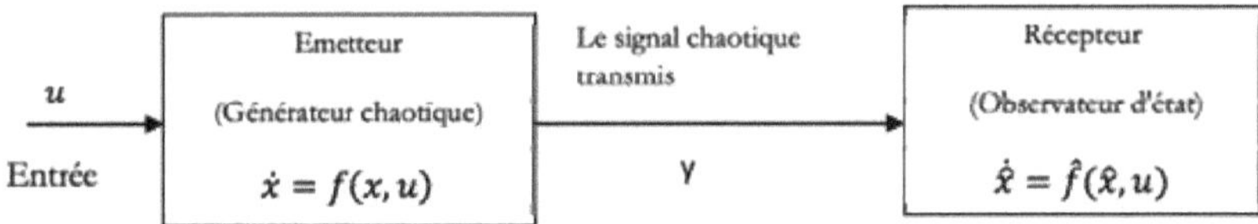

Fig.3.2-Principle of synchronisation using an observer.

For this principle, we say that the transmitter and receiver synchronise if the observer system $\dot{\hat{x}} = \hat{f}(\hat{x}, u)$ converges towards the system $\dot{x} = f(x, u),$ the synchronisation problem comes down to determining a function $\hat{f}$ such that :

$$\lim_{t \to \infty} \|x(t) - \hat{x}(t)\| = 0 \qquad (3.3)$$

3.2.2.2 Identical synchronisation

Identical synchronisation based on the coupling properties of two or more systems. Identical synchronisation developed on the basis of coupled chaotic circuits. To illustrate the method of synchronisation by coupling between two chaotic systems, we have chosen to present the identical synchronisation proposed by *Pecora* and *Carroll.*

3.2.2.3 Pecora and Carroll identical synchronisation

This synchronisation is based on the "Master-Slave" concept. The purpose of a Slave signal is to faithfully reproduce the Master signal. The "Master" system is also called the transmitter and the "Slave" system is called the receiver. Fig.3.3 shows the process of decomposition into sub-systems:

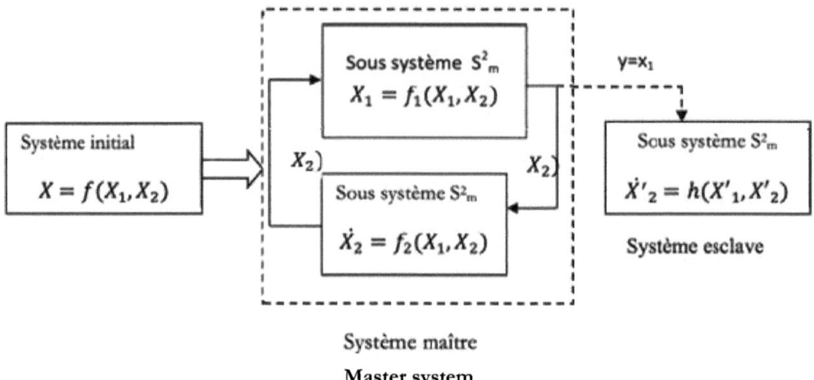

Système maître

Master system

Fig.3.3- Synchronisation structure using

decomposition into sub-systems.

Suppose we have an identical chaotic system of dimension n $\dot{X}(t) = f(X(t))$ is decomposed into two

subsystems $; X = f(X_1, X_2)$ such that

$$\begin{cases} \dot{X}_1 = f_1(X_1, X_2) \\ \dot{X}_2 = f_2(X_1, X_2) \end{cases} \qquad (3.4)$$

Of dimension (m and k) respectively with n-m+k.

Pecora and Carrol then derived from this system a subsystem X'_2 identical to subsystem X, such that

$\dot{X}'_2 = h(X_1, X'_2)$.

Pecora and *Carrol*'s configuration for identical synchronisation of chaos will be used.

$$\left. \begin{array}{l} \dot{X}_1 = f_1(X_1, X_2) \\ \dot{X}_2 = f_2(X_1, X_2) \end{array} \right\} \quad \text{système maitre} \qquad (3.5)$$

$$\dot{X}'_2 = f_2(X_1, X'_2) \} \quad \text{système esclave} \qquad (3.6)$$

master system

slave system

Note that the coupling between the systems is produced by the variable $X^\wedge$ in the master system (3.4), which is substituted for its analogue $X^\wedge$ in the subsystem (3.6). Thus the synchronisation of two outputs X_2 and X'_2 implies :

$$\Delta X_2 = \lim_{t \to \infty} \|X'_2 - X_2\| \to 0 \qquad (3.7)$$

This result can be checked as follows:

$$\Delta X_2 = X'_2 - X_2 \rightarrow \Delta \dot{X}_2 = f_2(X_1, X'_2) - f_2(X_1, X_2) \qquad (3.8)$$

Who is equal to

$$D_{X_2} f_2(X_1, X_2) \times \Delta X_2 \qquad (3.9)$$

Ou

$$D_{X_2} f = \left. \frac{\partial f_2}{\partial X_2} \right|_{\vec{X_2} = \vec{X_2}(t)} \qquad (3.10)$$

Thejacobian matrix of subsystem X_2

We deduce that the study of the convergence of ΔX_2 to zeros amounts to the study of the Lyapunov exponents of the subsystem X_2. As a consequence, Pecora and Carrol have stated the following theorem:

Theorem.3.1:

The master and slave systems are synchronised if and only if all the Lyapunov exponents of the slave system, called the conditional Lyapunov exponents, are negative.

3.5 Conclusion

In the first part of this chapter we presented the main synchronisation methods and then explained the principle of the identical synchronisation method by *Pecora* and *Carroll*. The synchronisation technique presented by *Pecora* and *Carroll* suffers from a high sensitivity to parameter variations. In order to obtain perfect synchronisation between the master and slave systems. What remains to be done is to find the appropriate control parameters, and this is what we are going to do by applying the optimisation algorithms we discussed earlier.

Chapter 4

Proposed methods for secure transmission of digital audio digital

4.1 Introduction

We will present a project for a high-security communication system using two levels of encryption based on chaotic systems. The first level is chaotic masking, while the second level is chaotic scrambling. The model of the proposed system is shown in Fig.4.1.

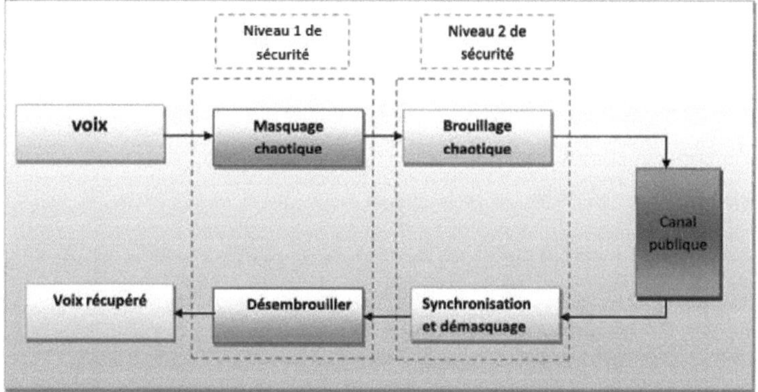

Fig.4.1 Block diagram of the proposed encryption system.

4.2 Encryption methods

At the masking level, we will choose two chaotic generators: the first chaotic generator is a hybrid between the *logistics map* (1.1) and the *tent map* (1.2), while the second generator is the *Chua* system. At the scrambling level, a scrambling key is used via the Arnold map (1.3) to broadcast signal samples using a secret key.

4.2.1Hiding in a chaotic generator: hybridisation between *logisticsmap* and *tent map*

4.2.1.1 Transmitter study

In this method, we present a new encryption scheme to improve the security of voice information in communication systems. We construct a hybrid of three approaches at the transmitter level:

♦   ♦♦ Chaotic maps: *Lxgpstique map* (1.1), *Tantmcp* (1.2) to generate an arbitration vector by some values mainly initiated to be join the original speech signal.

♦   ♦♦ A *watermarking* image is embedded in the encrypted signal for verification purposes, the watermarked audio is combined with the chaotic signal using a chaotic key to generate a file during the decryption process. The encrypted signal must be authentic and free from attack at the end.

♦   ♦♦ The third approach uses a scrambling key by Amold's map (1.3) is used to broadcast signal samples using a secret key, recovery of the original signal from the samples is not possible without this key. See figures Fig.4.2.

The encryption process begins by reading a voice signal stored on the hard disk using a Matlab function, where Matlab recommends representing the voice file in the range [-1 ,1], and then also reading the *watermark* file. *The* next steps are as follows:

1-   In this step, the user enters a key (key 1) to securely embed the watermark into the original speech signal. The scheme considered for embedding the watermark is shown in [62]. In this method, they embedded the *watermark* in the DCT and DWT domain and used a technical subsampling. This method

31

offers control of the embedding of the transparency and robustness of the watermark with an offset value (Δ). This step produces a speech signal marked with one of the secret information (*Watermark*) named Wtr_Sp.

2-   The *logistic map* (1.1) and the *tentmap* (1.2) create two chaotic signals, depending on the initial values entered by the user. These chaotic signals are generated and named Lg_S and Tn_S, respectively.

3-   Using the formulae below, the three signals Lg_S, Tn_S and Wtr_Sp are mixed to produce a new signal called Mx_Sg :

$$\begin{cases} Mx\_Sg_i = Tn\_S_i \times Wt\_Sp_i + (1 - Tn\_S_i)Lg\_S_i) - 1; & Wt\_Sp_i \geq 0 \\ Mx\_Sg_i = Tn\_S_i \times Wt\_Sp_i + (1 - Tn\_S_i)Lg\_S_i) + 1; & Wt\_Sp_i < 0 \end{cases} \qquad (4.1)$$

Where i: represents the sample index.

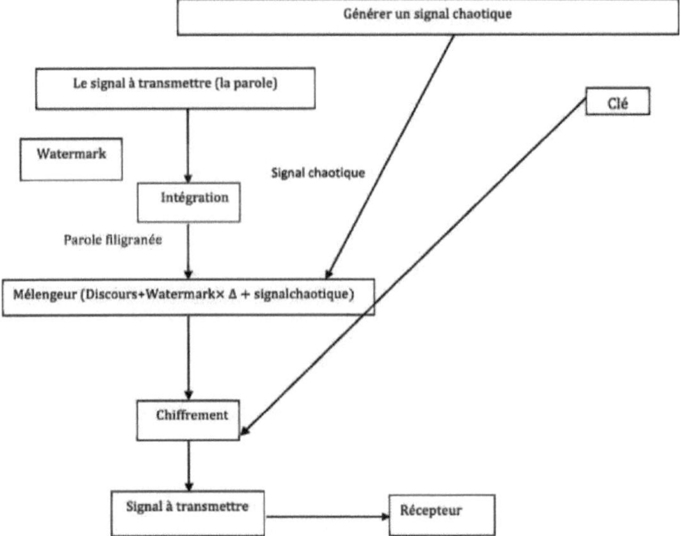

Fig.4.2 Encryption scheme flowchart

4-   Decompose the Mx_Sg into segments, where each segment length is a square number.

5-   Before applying the Amold transformation, each segment is reshaped into a 2D matrix (N XN elements).

6-   The user inserts another key (key 2), then the encryption process uses this key on each matrix to scramble its elements with the Amold transformation.

7-   Transform each scrambled matrix into an ID vector of length N2.

8-   To obtain the final encrypted voice signal, the encryption process collects the segment with another.

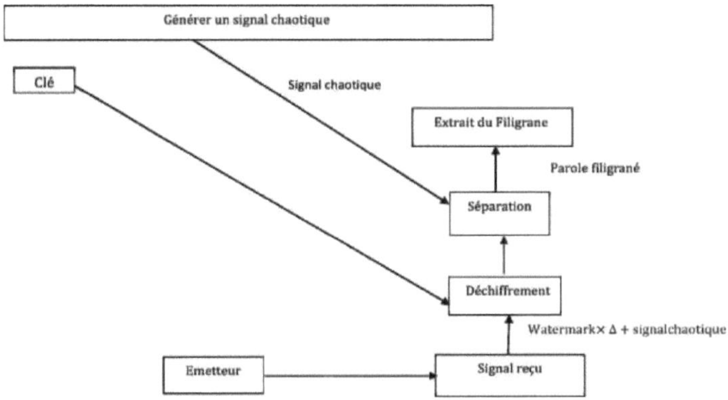

Fig.4.3. Flow chart of the decryption diagram

4.2.1.2 Study of the receiver

The receiver is used to retrieve the message (voice signal). Using the previous chaotic key, the received signal is decrypted by deleting the same chaotic signal generated at the transmitter. We extract the watermark from the decrypted signal and check the signal obtained against the original to ensure its originality without degradation, see Fig.4.3 :

1-   Steps 4 and 5 of the encryption process are applied to the encrypted voice signal

2-   The inverse Arnold transform is applied to each 2D matrix using the same key (key 2) as before.

3-Re-model each recovered matrix as an ID vector of length N2.

4-Combine the recovered segments with each other to produce Mx_Sg_i'.

5- The same second stage of the encryption process is applied without modifying the initial value.

6-The separation of samples of decrypted voice signals is carried out in compliance with the following elements

$$\begin{cases} Wtr\_Sp_i' = \dfrac{Mx\_Sg_i' + 1 - (1 - Tn\_S_i)Lg\_S_i)}{Tn\_S_i} & Mx\_Sg_i < 0 \\[2mm] Wtr\_Sp_i' = \dfrac{Mx\_Sg_i' - 1 + (1 - Tn\_S_i)Lg\_S_i)}{Tn\_S_i} & Mx\_Sg_i \geq 0 \end{cases} \qquad (4.2)$$

Where: $Wtr\_Sp'_i$ is the decoded speech signal and $Mx\_Sg'_i$ is the result of the fourth step.

Up to this stage, the voice signal is decrypted, but not confirmed. To check that the voice signal is safe, sent on the authentication side, the decryption process is maintained with these steps:

7-Extract the continuous watermark (*Walermark*) in the encrypted speech signal using the same key (del) in the extraction process given in [62].

8-The authentication of the decrypted voice signal is controlled by verifying the similarity between the extracted watermark and the original watermark. Done, the more similarity between the two decrypted voice means the more authenticated.

4.2.2 Masking by a chaotic *Chua* system

In this section, we present a proposal for a high-security communication system using two levels of encryption based on high-dimensional (HD) chaotic systems. At the transmitter end, the first level is chaotic masking using the *Chua* chaotic system, while the second level is scrambling of the encrypted audio using an *Alrond map*. At the receiver, the masking is removed using a *Pecora Carroll* synchronisation version of the chaotic system.

4.2.2.1 Transmitter study

The transmitter is a continuous-time chaotic system with the following *Chua* circuit (1.4):

$$\begin{cases} \dot{x}_1 = \alpha(x_2 - x_1 - f(x_1)) \\ \quad \dot{x}_2 = x_1 - x_2 + x_3 \\ \quad \dot{x}_3 = -\beta(x_2 - x_3) \end{cases} \tag{4.3}$$

To mask the original audio signal after accessing the initial conditions and the parameters of the *Chua* system respectively: $x_1(0) = 1, x_2(0) = 0.5, x_3(0) = -1$ and $\alpha = 15.7 \text{et} \beta = 28,$ which are considered as an encryption key (del), Fig.4.5 represents the chaotic masking model. Details of the chaotic masking and scrambling algorithm are given below:

1-   The audio signal m(t) is added to a chaotic chua generator control signal to obtain an encrypted audio signal *s(t)*.

2-   The encrypted audio signal is divided into frames of known length, each of which is divided into sub-frames (segments).

3-   Before applying the Amold transformation, each segment is remodeled into 2-D matrix (NXN) elements.

4-   After inserting another key (key 2), the process uses this key on each matrix to scramble these elements by extending Arnold's algorithm.

5-   Transform each scrambled matrix into an ID vector of length N2.

6-   To obtain the final encrypted voice signal, the encryption process collects it wisely with another.

To recover the encrypted audio signal we will synchronise the transmitter and receiver using *Pecora* and *Carroll* synchronisation.

4.2.2.2 Pecora and Carroll synchronisation of two Chua chaotic systems

We use the circuit in Fig.1.6 as the transmitter, after adjusting the R value to obtain the chaotic regime, and the circuit in Fig.4.4 as the receiver. We adjust the control of the receiver parameter R to also obtain the chaotic regime. Both circuits are then able to operate in their chaotic mode (double rolling). Coupling is achieved between the two circuits by means of the electrical voltage V_{CI} from the transmitter. This signal from the transmitter passes through an operational amplifier follower before being used to couple to the receiver r(t).

The receiver is divided into two sub-systems, clearly shown in Fig.4.4. These two sub-systems are linked together by an operational amplifier follower to decouple the two sub-systems. The first subsystem comprises the capacitor C_2 , the resistor $R,$ and the inductance $L.$ The second sub-system consists of resistance R, capacitance C_I and *Chua* resistance N_R .

The behaviour of the first RL subsystem of the receiver is given by the following system of equations:

$$\begin{cases} C_2 \dfrac{dV_2}{dt} = \dfrac{1}{R}(r(t) - i_L) \\ \quad L \dfrac{di_L}{dt} = -V_2 \end{cases} \tag{4.4}$$

The voltage V_1 controls the second RC subsystem. Its behaviour is
is given by the following equation:

$$C_1 \frac{dV_1}{dt} = \frac{1}{R}(V_2 - V_1) - f(V_2) \tag{4.5}$$

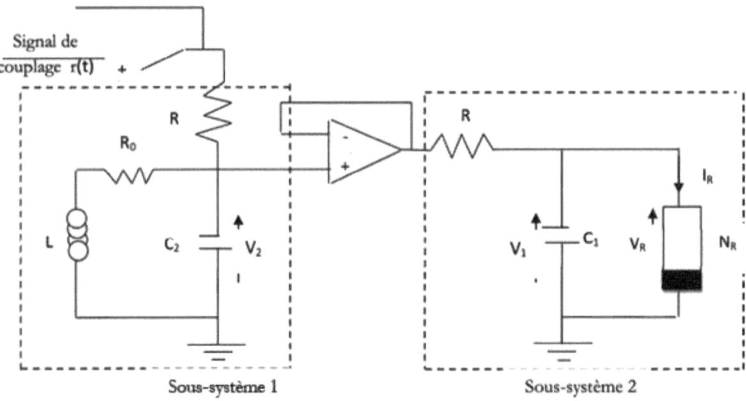

Fig.4.4 Chua receiver divided into sub-systems.

When the values of the electronic components of the Chua resistance are the same, we observe a synchronisation phenomenon between the transmitter and the receiver. The voltage V_1 at the receiver synchronises with the voltage V_1 at the transmitter. They show the same temporal variations.

4.2.2.3 Receiver study

The receiver is a system identical to the transmitter plus a simple subtractor to successfully remove chaotic masking, according to the *Pecora* and *Carroll* synchronization scheme as explained earlier. This scheme is based on sending a control signal X_m, which is a simple addition between the transmitter output signal $y(t)$ and the message $m(t)$. The signal s(t) is transmitted to the receiver through the transmission channel, assuming that only this channel is ideal, see Fig.4.5.

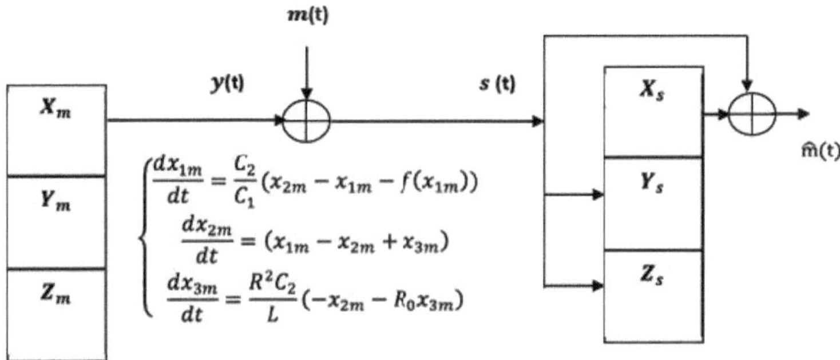

$$\begin{cases} \dfrac{dx_{1m}}{dt} = \dfrac{C_2}{C_1}(x_{2m} - x_{1m} - f(x_{1m})) \\ \dfrac{dx_{2m}}{dt} = (x_{1m} - x_{2m} + x_{3m}) \\ \dfrac{dx_{3m}}{dt} = \dfrac{R^2 C_2}{L}(-x_{2m} - R_0 x_{3m}) \end{cases}$$

Fig.4.5 Chaotic masking and demasking of the *Chua* system.

4,2,2,4 The control signal for synchronisation

Or the following Maitre du *Chua* system:

$$\begin{cases} \dfrac{dx_{1m}}{dt} = \alpha(x_{2m} - x_{1m} - f(x_{1m})) \\[2mm] \dfrac{dx_{2m}}{dt} = (x_{1m} - x_{2m} + x_{3m}) \\[2mm] \dfrac{dx_{3m}}{dt} = \beta(-x_{2m} - R_0 x_{3m}) \end{cases} \qquad (4.6)$$

For the choice of the slave subsystem, three configurations of two variables are possible

$$\begin{cases} \dfrac{dx_{1s}}{dt} = \alpha(x_{2s} - x_{1s} - f(x_{1s})) \\[2mm] \dfrac{dx_{2s}}{dt} = (x_{1s} - x_{2s} + x_{3s}) \end{cases} \qquad (4.7)$$

(x_{1m}, x_{2m}) configuration with x_{3m} coupling inlet

$$\begin{cases} \dfrac{dx_{1s}}{dt} = \alpha(x_{2s} - x_{1s} - f(x_{1s})) \\[2mm] \dfrac{dx_{3s}}{dt} = \beta(-x_{2s} - R_0 x_{3m}) \end{cases} \qquad (4.8)$$

(x_{1m}, x_{3m}) : configuration with x_{2m} coupling inlet

$$\begin{cases} \dfrac{dx_{2s}}{dt} = (x_{1s} - x_{2m} + x_{3s}) \\[2mm] \dfrac{dx_{3s}}{dt} = \beta(-x_{2m} - R_0 x_{3s}) \end{cases} \qquad (4.9)$$

Configuration (x_{2m}, x_{3m}) : with x_{1m} coupling inlet

In our work we propose the Master system (4.6) with the Slave system (4.9) as shown in Fig.4.6.

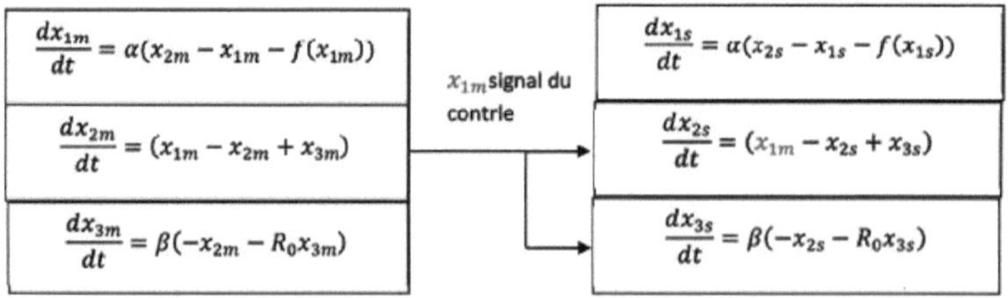

Fig.4.6. Synchronisation method by *Pecora and Caroll.*

To do this, a set of vectors called state error vectors, representing the difference between the master and slave states, is defined.

The error states e_{x1}, e_{x2} et e_{x3} are given by

$$e_{x1} = x_{1m} - x_{1s}$$

$$e_{x2} = x_{2m} - x_{2s} \qquad\qquad (4.10)$$

$$e_{x3} = x_{3m} - x_{3s}$$

Subtracting the master system from the slave system gives :

$$\dot{e}_{x1} = \dot{x}_{x1m} - \dot{x}_{x1s}$$
$$= (a(x_{2m} - x_{1m} - f(x_{1m}))) - (a(x_{2s} - x_{1s} - f(x_{1s}))) \qquad\qquad (4.11)$$
$$= a(x_{2m}\text{-}x_{2s} - f(x_{1s})) - (x_{1m}\text{-}x_{1s}) - f(x_{1m}))$$
$$= a(e_{x2}\text{-}e_{x1})$$

And we also get :

$$\dot{e}_{x2} = e_{x1} - e_{x3} \qquad\qquad (4.12)$$

$$\dot{e}_{x3} = -\beta(e_{x1} + R_0 e_{x3}) \qquad\qquad (4.13)$$

Which can be described as :

$$\dot{e} = A(t)e \qquad\qquad (4.14)$$

It can be shown that the synchronization complete truth is globally asymptotically stable at the origin for any choice of control signal for these conditions.

Considering:

$$E(e) = 1/2(1/a e_{x1}^2 + e_{x2}^2 + 1/\beta e_{x3}^2) \qquad\qquad (4.15)$$

$$\dot{E}(e) = (1/a \dot{e}_{x1} e_{x1} + \dot{e}_{x2} e_{x2} + \frac{1}{\beta} \dot{e}_{x3} e_{x3})$$

$$\qquad\qquad (4.16)$$

$$= ((e_{x2} - e_{x1})e_{x1} + (e_{x2} - e_{x3})e_{x2} - (e_{x2} + R_0 e_{x3})e_{x3}))$$

$$= -e_{x1}^2 + e_{x1} * e_{x2} + e_{x2} * e_{x2} - e_{x2}^2 - e_{x3} * e_{x2} - R_0 e_{x3}^2$$

$$= -\left(e_{x1} - \frac{1}{2}e_{x2}\right)^2 - \frac{3}{4}e_{x2}^2 - R_0 e_{x3}^2$$

We know that $R_0 > 0$, and from $\dot{E} < 0$, , done and according to the direct lyapunov exponent method the system is asymptotically stable.

Therefore, whatever the initial conditions imposed between the transmitter and receiver systems, synchronisation occurs when $t \to \infty$:

$$\lim_{t\to\infty} e_{x1}(t) = \lim_{t\to\infty} e_{x2}(t) = \lim_{t\to\infty} e_{x3}(t) = 0 \qquad\qquad (4.17)$$

Message (audio) recovery depends on the synchronisation error; the smaller the synchronisation error, the greater the recovery.

Returning to Fig.4.5, the signal received is given by :

$$s(t) = X_m(t) + m(t) \qquad\qquad (4.18)$$

and the recovered speech signal is:

$$\widehat{m}_s(t) = s(t) - X_s(t) = [X_m(t) + m(t)] - X_s(t) = e(t) + m(t) \qquad (4.19)$$
$$\widehat{m}_s(t) = e(t) + m(t)$$

With $\quad X_m(t) - X_s(t) = e_{x1}(t)$

The effect of the presence of the information signal during synchronisation at the receiver is taken into account. Ameer. K, Jawad et al. have shown that to neutralise this effect, the information signal is sent back to the chaotic transmitter [17]. In other cases, the synchronisation error is caused by variations in the transmitter and receiver parameters, as it is not very easy to make the two systems similar.

4.4 Evaluation of the methods proposed in our work:

In the encryption process, three types of test are carried out to verify the performance of the proposed systems. These tests are:

♦♦♦ the verification of cryptanalysis and the verification of the intelligibility of secure speech or testing the quality of encryption, as well as the control and authentication of *watermarks:*

4.4.1 Evaluation of the encryption method using chaotic maps *(logistics, tent, Arnold)*

4.4.1.1 Chaotic cryptanalysis :

Cryptanalysis aims to find key space and sensitivity

-Espace de cle:

The keys in this method are composed of the initial conditions and parameters $(a_0 , r\sim)$, (b_0 , u) of logistic map and tant map respectively as shown in tab 5.2, and the scrambling key by *Arroldmap,* we will compute the key space from these keys.

-Key sensitivity analysis

We have also tried to test and examine the sensitivity of the encryption algorithm by changing one or more keys. This was done by calculating the NSCR and the UACI.

4.4.1.2 Verifying the unintelligibility of speech

This is an important factor in the process of masking and scrambling speech.

- Effect of encryption and dëcryption processes:

We will measure the SNR values and the value of the correlation coefficient in the case of encryption and decryption.

4.4.1.3 *Watermark* control and authentication :

The purpose of this operation is to further enhance security and credibility, and to resist certain AWGNs which add Gaussian white noise to BER values.

4.4.2 Evaluation of the encryption method using the *Chua* and *Arnoldmap* chaotic system

We will evaluate the effectiveness of this method using the same coefficients as above, which have the initial values and the parameters $(x_{1m0}, x_{2m0}, x_{3m0}, \alpha, \beta, K)$ of the *Chua* and *Arnold* circuits respectively.

4.5 Conclusion

In this chapter, we have proposed a new scheme for the secure transmission of a voice signal based on chaotic systems. The proposed scheme uses two levels of encryption based on chaotic systems: the first level is masking, while the second level is scrambling. In the masking level, we use two types of system: in the first, we have masked the audio using a hybrid between logistic maps (*map* and *tent*), while in the second method we have masked the audio signal using the *Chua* chaotic system. In the latter case, the message will be recovered in the receiver after the *Pecora* and *Carroll* synchronisation.

General conclusion

General conclusion

With the development of modem communications and multimedia technologies, and because of the sensitivity of voice data, it has become very important to protect this data over digital communications with fast and secure encryption systems before transmission or distribution without worrying about interception and its sensors. To this end, we have designed two high-security voice communication systems using two levels of encryption based on the chaotic systems described in Chapter 1. The first level is chaotic masking, which is based on two types of chaotic systems, while the scrambling level uses the *Arlond* map *(Cat map)*.

In the first method, we used a hybridization of the chaotic maps *(logistic map and tent map)* with the original audio signal. Integrating the low-dimensional chaotic system optimizes the key space, and results in a complex system for integrating the original values of the speech signal, but this leads to higher computational complexity and computation time.

In the second approach, we have masked audio using a *Chua* chaotic system which is one of the systems with rich and complex dynamic behaviours and a large key space. Recovering the confidential audio first involves synchronising the two chaotic systems and using *Pecora and Caroll,* which is described in detail in Chapter 3. *Chua*'s chaotic system gives high security with a larger key space and real time operation, because of the sensitivity of the keys to a slight change in one of the parameters or in the initial state of *Chua*'s system and unexpected signal state values.

References

[1]   D. Ambika. And V. Radha, "Secure Speech communication - A Review", International Journal of Engineering Research and Applications (IJERA), Vol. 2 Issue 5 PP. 1044-1049 (2012).

[2]   B. Sadkhan SattarandA. Abbas Nidaa, "Performance Evaluation of Speech Scrambling Methods Based on Statistical Approach" ATTI DELLA "Fonazione Giorgio Ronchi" AnnoLxvi, No. 5 PP. 601-6014 (2011).

[3]   R. Gnanajeyaraman, K. Prasadh and Dr. Ramar, "Audio encryption using higher dimensional chaotic map", International Journal of Recent Trendsin Engineering, Vol. 1, No. 2, pp. 103- 107,May2009

[4]   M. Ahmad, B. Alam and O. Farooq, "Chaos based mixed key stream generation for Voice data encryptions", International Journal on Cryptography and Informaton Security (IJCIS), Vol. 2, No. 1, March 2012.

[5]   A. V. Prabu, S. Srinivasarao, T. Apparao, M. J. Rao and K. B. Rao, "Audio encryption in handsets", International Journal of Computer Applications, Vol. 40 ,No. 6, pp. 40-45, Feb. 2012

[6]   M. Ashtiyani, P. M. Birgani and S. K. Madahi, "Speech Signal Encryption Using Chaotic Symmetric Cryptography", Journal of Basic and Applied Scienctific Research, Vol. 2, No. 2, pp. 1668-1674, 2012

[7]   F. Anstett, "Les systemes dynamiques chaotiques pour le chiffrement: synthese et cryptanalyse", These, Hanri Poincare University, Nancyl, 2006

[8]   E. Cherrier, "Estimation de I'etat et des entrees inconnues pour une classe de systeme non lineaires", These, Institut National Polytechnique de Lorraine, 2006.

[9]   S.Najim Al Saad, E. Hato, "A Speech Encryption based on Chaotic Map", International journal ofComputer Application (0975-8887), Vol. 93, No. 4, May 2014.

[10] M. Ammar Raheema, B. Sattar, S. SMIEE, M Sinan Majid , "Performance Enhancement of Speech Scramling Technique Based on Many Chaotic Signal", International Conference on Computer Science and Engineering (CSASE), Duhok, Kurdistan Region-Iraq,2020. October 2015.

[11] E. Hato, S. Dayla, "Lorenz and Rossler Chaotic System for Speech Signal Encryption", International Journal of Computer Application (0975-8887), Vol. 128, No. 11,

[12] F. Mahmode, M. Shalaby, Y. Kamal, S. El Ramly, "A Speech Cryptosystem Based on Chaotic Modulation Technique", Egyptionjournal ofLanguage Engineering, Vol. 4,No,l,2007.

[13] N. Hikmat Abdullah, S. Saad. Hreshee, and K. Ameer "Design of Efficient Noise Reduction Scheme For Secure Voice Masked By Chaotic Signals", Journal of American Science 2015; Vol. 11, No.7, pp. 49-55.

[14] K. Ameer, N. Abdullah, S. Saad /Secure Speech Comminication System Based on Scrambling and Masking by Chaotic Map", International Conference on Advance in Sustainable Engineering and Application (ICASEA), Wasit University, kut, Iraq, 2018.

[15] U. Parlitz, L.O. Chua, L. Kocarev, K.S. Halle and A. Shang, 'Transmission of digital signals by chaotic synchronization", International journal of Bifurcation and Choas, Vol. 2,No. 4, pp. 973-977, 1992.

[16] L.M. Pecora and T.L. Carroll, "Synchronization in Chaotic Systems", Physicals Review and Letters, pp. 821-824,1990.

[17] L.M. Pecora and T.L. Carroll, 'Synchronization Chaotic Systems", *IEEE Trans. Circuitand Systems*, vol. 38, pp. 453-456, 1991

[18] M. Lakshmanan, S.Rajaseekar, "Nonlinear Dynamics Integrability, Chaos and Patterns". Advanced Texts in Physics, Publisher Springer-Verlag Berlin Heidelberg, 2003.

[19] S. Wiggins, "Introduction to Applied Nonlinear Dynamical Systems and Chaos", Texts in Applied Mathematics, Springer-Verlag New York, 2003.

[20] A. Wolf, J.B. Swift,H.L.Swinneyand J.A.Vastano, "Determining Lyapunov exponents from a time series", Physica D, Vol. 16, pp. 285-317, 1985.

[21] M. L'Hemault, "Feasabilite d'un Systeme d'Emission-Reception Analogique pour les Communications Securisees parle Chaos", Tese, Universite de CergyPontoise, 2007.

[22] L.O. Chua, C.W. Wu, A. Huang and G.-O. Zhong, *"K universal Circuit for Studying and Generating Chaos-Part I: Routes to Chaos'IEEE Trans. Circuits and Systems-!: fundamental Theory and Application,* Vol. 40, No. 10, October, 1993.

[23] L*.O. Chua, 'Global Unfolding of Chua's Circuit", *IEEE Trans. Fundamental,* Vol. 76, No. 5, pp.704733, 1993.

[24] L.O. Chua, L. Kovarev, K. Eckert, and M. Itoh, *"Experimentalchoas synchronisation in Chua's circuit",* International journal ofBifurcation and Choas, Vol. 2,pp. 705-708,1992.

[25] E. Ott, C. Grebogi,&Yorke, J. A. [1990] "Controlling chaos," PhysicalReviewLetters64, pp. 1196-1199.

[26] L.O. Chua, & G.N. Lin, [1990] "Canonical realization of Chua's circuit family," IEEE Transaction on Circuits and Systems-I 37, pp. 885-902.

[27] A. V. Prabu, S. Srinivasarao, T. Apparao, M. Jaganmohan, and K. Babu Rao, "Audio encryption in handsets," *international\journal ofComputer.Applications,* vol. 40, no. 6, February 2012

[28] B. Boulebtateche, M. M. Lafifi, and S. Bensaoula. A multi media chaos-based encryption algorithm. [Online]. vailable:https://www.researchgate.net/publication/228437181.

[29] G. Djamal Eddine , "Logistic function and chaotic standard for image encryption satellitaires",thesis majistere,2011, Universite Mentouri de Constantine

[30] R.James Drummond, **"Analogue-to-Digital** Conversion",Microprocessor Interfacing Techniques, PP.89-108, September 1997.

[31] http://www.tsp.ecemcgil.ca/mmsp/documents / AudioFormats/

[32] Brooks,W. David ; Carr, Adam and Edkins, Keith; et al, (2004), "Data Compression", Wikipedia the Free Encyclopedia, GNU Free Documen -tation License, Boston, U.S.A.

[33] A. Chan Carusone, "Digital Algorithms for Analog Adaptive Filters", Ph.D. Thesis, University of Toronto, 2002.yption. IEEE Trans. Image Process. 15, 2061-2075 (2006).

[34] M..tahon, "signal processing",Acoustics laboratory. Conservatoire National des Arts et Metiers, 2014-2015.

[35] A. kadhim Jawad, "Design and Simulation of Secure Communication System based on Chaos overAWGN Channel",Al-Mustansiriya University,2015

[36] U. Parlitz, L.O. Chua, L. Kocarev, K.S. Halle and A. Shang, *"Transmission of digital signals by chaotic synchronisation"*, International Journal of Bifurcation and Choas, Vol. 2,No. 4, pp. 973977, 1992.

[37] T. Yang, C. Wah-Wu and L. Chua, 'Cryptography Based on Chaotic Systems'*IEEE Trans, Circuit and Systems-!: FundamentalTheory andApplication,* Vol. 44, No. 5, pp. 469-472, May 1997.

[38] F. Anstett, 'Les systemes dynamiques chaotiques pour le chiffrement: synthese et cryptanalyse", These de doctorat, Université de Henri Poincare, Nancyl, 2006.

[39] T. Yang and L. Chua ,'Secure Communication via Chaotic Parameter Modulation", *IEEE Trans. Circuits and Systems-!: fundamental Theory andApplications,* Vol. 43, No. 9, pp. 817-819, September 1996.

[40] R. Channapragada Seshagiri Rao, Munaga V.N.K. Prasad "Digital WatermarkingTechniques in Curvelet and Ridgelet Domain", Springer Briefs in Computer Science (2016), DOI 10.1007/978-3-31932951-2.

[41] M. Hemis ,"Systeme de Tatouage pour la Securite des Données audio ", These de Doctorat , Universite des sciences et technologies de houari Boumediene ,2017.

[42] S. Slami, A. Merrad, A. Benziane, Novel secured scheme for blind audio/speechnorm-spacewatermarking by Arnold algorithm, SignalProcessing154   (2019)   74-86, www.elsevier.com/locate/sigpro.https://doi.Org/10.1016/i.sigpro.2018.08.011

[43] A. Merrad, S. Slami, "Blind speech watermarking using hybrid scheme based on DWT/DCT and sub-sampling", *Midtimed ToolsAppl*(2018) 77:27589-27615, https://doi.org/10.1007/sllQ42-018-5939-z

[44] A. Merrad, S. Slami, A.Benziane, A. Hafaifa, Robust Blind Approach for Digital Speech Watermarking, 2018 2nd International Conference on Natural Language and Speech Processing (ICNLSP), 978-1-5386-4543-7/18IEEE. DOI: 10.1109/ICNLSP.2018.8374366.

[45] H. Delfs, H.Knebl, "introduction to cryptography", Springer verlag, Berlin, 2002.

[46] A. Kerckhoffs, "La Cryptographic Militaire". Journal Des Sciences Militaires. Vol.9, pp .238.161-191, 1883

[47] H. Khanzadi, M.Eshghi and Shahram EtemadiBorujeni "Image Encryption Using Random Bit Sequence Based on Chaotic Maps", Arab J SciEng (2014) 39:1039-1047, DOI 10.1007/sl3369-013-0713-z

[48] C.K. Huang, H.H. Nien "Multi chaotic systems based pixel shuffle for image encryption", Optics Communications 282 (2009) 2123-2127, doi:10.1016/j.optcom.2009.02.044.

[49] H. Liu, B. Zhao and Linquan Huang "Quantum Image Encryption Scheme Using Arnold Transform and S-box Scrambling", Entropy 2019, 21, 343; doi:10.3390/e21040343.

[50] P. Sathiyamurthi and Ramakrishnan /'Speech encryption algorithm using FFT and 3DLorenz-logistic chaoticmap", Multimedia Tools and Applications(2020),https://doi.org/10.1007/sll042-020-08729-5.

[51] B. Ratner 'The correlation coefficient: Its values range between + 1/ -l,ordo they ?', Journal ofTargeting, Measurement and Analysis for Marketing (2009) 17,139 - 142. doi: 10.1057/jt.2009.5

[52] FJ.Farsana , K. Gopakumar "A NovelApproach for Speech Encryption: Zaslavsky Map as Pseudo Random Number Generator", 6th International Conference on Advances In Computing&

Communications, ICACC 2016, 6-8 September 2016, Cochin, India (Procedia Computer Science 93 (2016) 816 - 823).

[53] K. Gopakumar, FJ. Farsana and V.R. Devi "An audio encryption scheme based on Fast Walsh Hadamard Transform and mixed chaotic key streams" Applied Computing and Informatics (Published by Emerald Publishing Limited 2019), D01:10.1016/j.aci.2019.10.001

[54]]Y.S. Tang, A.I. Mess and L.O. Chua, "Synchronization and chaos*", IEEE Trans. Circuit and Systems*, Vol. 30, pp. 1-2, 1983.

[55] L.M. Pecora, and T.L. Carroll, "chaotic circuits," lEEEtrans Circuits yst.
vol. 38, pp. 453-456, 1990.

[56] L.M. Pecora, and T.L. Carroll, "Driving systems with chaotic signals," Phys. Rev. A44, pp. 2374-2383, 1991.

[57] H. Hamiche/'Inversion a gauche des systèmes dynamiques: Application a la transmission securisee des données' ,These de doctorat, Universite moumoudmammeri de tizi ouzou,2011.

[58] H. Nijmeijer and Iven M.Y Mareels, "An observer Looks at synchronization", *IEEE Trans. Circuit systems: Vundamcntal Theory and Application,* vol. 44, October 1997.

[59] G. Kreisselmeier, Adaptative observer with exponential rate of convergence, IEEE, Transactions on Automatic and Control, vol, 22.pp. 2-8 ,1977.

[60] Y. Xiong and M. Saif, "Sliding Mode Observer for Nonlinear Uncertain Systems",*IEEETrans. Automatic Control,* Vol. 46, pp. 2012-2017, No. 12, December 2001.

[61] A. Maybhate and R.E. Amritkar, "Use of synchronization and adaptive control in parameter estimation from a time series," *TPysicalEeviewE,* Vol. 59, 1999 pp. 284-293.

[62] S. Slami, A.Merrad, A. Benziane, "Novel secured scheme for blind audio/speechnorm- space watermarking by Arnold algorithm", *Signal Processing154* (2019) 74 86,www.elsevier.com/locate/sigpro. https://doi.org/10.1016/j.sigpro.2018.08.011.

Printed by Books on Demand GmbH, Norderstedt / Germany